Help for the Technology Shy presents...

The Little Computer Help Book

Michael Gorzka (Writer)

Mariana Miralles (Illustrator)

Henry Viera (Illustrator)

Tanya Zeinalova (Illustrator)

www.technologyshy.com

ISBN: 978-1-7349815-3-7

The author and publisher advises readers to take full responsibility for their technology use and have neither liability nor responsibility to any person or entity with respect to any loss or damages arising from the information contained in this book.

Table of Contents

Let's Talk Rocket Science 🚀 (Or Not 🙂)

Getting stuff done with computers is not even **close** to being rocket science.

In subsequent chapters, we will casually but carefully cover the thoughtful and methodical use of the *computer basics* in order to reach a clearly defined goal or objective — which is the magic formula for productive, stress-free contemporary computer use.

We'll talk about how to apply real world common sense to our computer use.

Painting in broad strokes

In order to keep things as simple as possible and to make this book timeless and applicable to **everybody** (regardless of which computer they are using and what they would like to do with said computer) your Tech Wizard Tour Guide will be teaching you how to fish instead of simply handing you a fish.

(while preparing you to expect the unexpected)

It's showtime!

How-to videos for everything we cover in this book ➕ tips & tutorials, additional resources and some fun stories await you on the Help for the Technology Shy website.

Please see the "Where to Go From Here" chapter for details.

use of italics

I Any *italicized* words or phrases you see in this book will be indexed in this book **and** subjects of Help for the Technology Shy how-to videos and short articles (i.e. blog posts).

Okay? Okay! Let's proceed!

Get a Technology Notebook

Documentation is the key, the key, the key. To be productive with your computer, you will need to keep track of things. You will need to keep on top of things.

In a nutshell, you will need an organizational system to record your passwords and other info related to your computer and your computer use.

A technology notebook will help keep you grounded and in control of your computer use.

It will be your rock & anchor and your boon companion before, during and after your computer use.

Make your technology notebook your bedside reading. Review it at least once a week.

Do not trust your memory.

For example, when you need to create a username & password combo, write it down right there and then.

Always proceed **methodically** and never try to rush through anything (especially your passwords). If you do, whatever you ran pell-mell through may come back at a later date and bite you.

Here are examples of things you should record into your technology notebook:

- Your computer's *login* password or code

- Your computer's *administrator* user name & password

- Your *cloud account* username and password (if you have one)

- *Wi-Fi passwords* (i.e. the ones that do not change on a daily or hourly basis)

- *Web addresses*

And always take the time to truly understand what something is.

For example, many of my technology shy friends have had some misconceptions about their computer's *login password* and their computer's *administrator password*.

Also document any questions you have about your computer use as they arise or pop into your head — and then get some *good help*.

"How do I share my contacts between my computer and my smartphone?"

Tips for using your technology notebook…

1. Keep it up to date. You should periodically change your passwords — and the 'old' passwords in your technology notebook should be removed and the new passwords carefully entered.

2. Keep it as neat as possible. If your handwriting is not neat (mine is the pits), use one page per entry and write in large letters.

3. Make your technology notebook your bedside reading until you feel calm, cool, relaxed and focused during your computer use — and then, review it weekly.

4. Don't lose it! If you take your technology notebook out of your

cloistered domicile, keep very close
tabs on it.

5. Take photos of its pages to have as a
back up; or perhaps photocopy it.

Contemporary technology use can be a
maze — documentation will be your
secret weapon.

Breathe

You need to be *cool as a cucumber* in order to stay focused as you are methodically following visual cues en route to completing your current task at hand.

Mindful Breathing

Mindfulness meditation involves focusing on your breathing and bringing your attention to the present without allowing your mind to drift off to the past or future.

1. Choose a calming focus, including a sound ("om"), positive word ("peace"), or phrase ("breathe in calm, breath out tension") to repeat silently as you inhale or exhale.

2. Let go and relax. When you notice your mind has drifted, take a deep breath and gently return your attention to the present.

Ask your local librarian for books on how to stay calm during stressful situations.

Press Your Emotional Reset Button

Computer use can be very intense!

Productive, stress-free computer use is roughly 90% mental acuity and 10% mouse clicks & keyboard presses.

This being the case, we can't have any past computer-related conundrums, calamities and/or miscues weighing you down, emotionally.

When you press your reset button, you **instantly distract your brain from its undesirable state of arousal,** and you require it to pay attention to this new physical stimulus. In the process, it begins to forget what it was doing. In a sense, you derail the signals that activate your amygdala - your emotional fire alarm.[1]

Okay, so please form a mental image of pressing your emotional reset button or bang a physical gong if you have one or ring your own doorbell and then have some ice cream — however you wish to carry out this important task.

---

[1] https://www.psychologytoday.com/us/blog/brainsnacks/201609/how-install-your-emotional-reset-button

Whatever you are going to do, do it now. We'll be here when you get back.

Welcome back and congratulations! You now have a fresh start. You have a clean slate.

Now let's continue your guided journey to computer confidence, calmness and proficiency.

How To Brew the Purrfect Cup of Tea

TEA BREWING PROCESS

1. Bring fresh, cold water to a rolling boil
2. Play a teabag in the cup
3. Add three-quarter cup of hot water
4. Patiently let it steep for 3 to 5 minutes
5. Add milk and sugar to taste
6. Stir with spoon
7. Enjoy!

Question: What does brewing a perfect cup of tea have to do with becoming comfortable and proficient with technology?

Answer: Everything, really.

When you make a cup of tea, you have a desired goal, an objective, a specific outcome in mind.

And to realize this objective, you methodically and mindfully follow a sequence of well-reasoned steps.

For example, you cannot pour the tea, add whatever you like to it and **then** heat the water — it wouldn't make logical sense to do so 🤭

Successful tea brewing also involves patience (3 - 5 minutes for steeping) and realistic expectations (e.g. your tea will not be hot unless you are willing to heat some water — however you choose to do it ♨️).

If you apply this same conceptual procedure to your everyday computer use then you're golden, you're good to go, you're off to the races 🏇 (figuratively speaking of course).

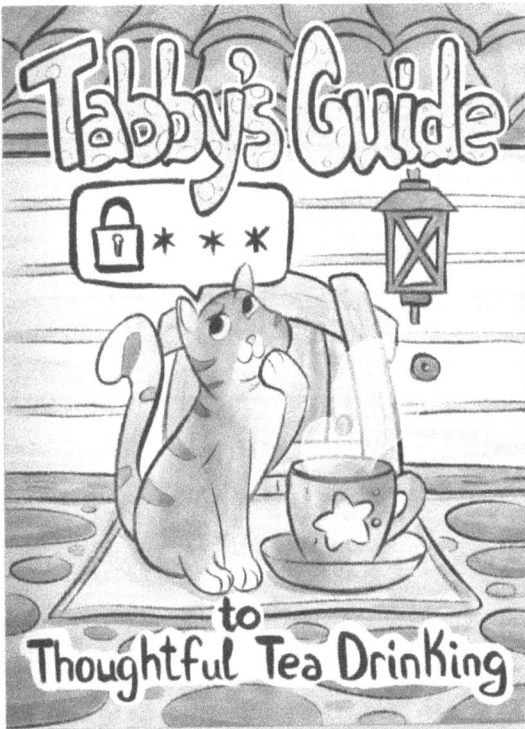

This is the only passwords book in existence (as far as I know) that doubles as a cat fancier's tea drinking journal.

Gaze at Your Computer's "Gooey"

If your computer is powered on, and it is working as it should, you will see a *graphical user interface.*

graph·i·cal us·er in·ter·face

noun COMPUTING
noun: GUI; plural noun: GUIs
a visual way of interacting with a computer using items such as windows, icons, and menus.

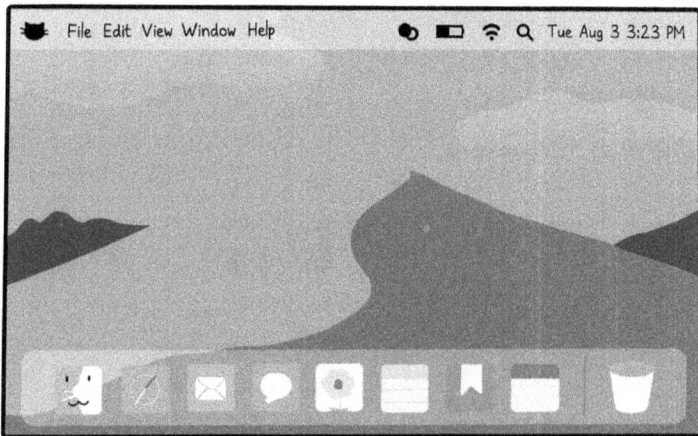

Abstract computer desktop showing icons, menus & a nifty background "wallpaper".

The paperclip ✑ and camera 📷 icons in the following illustration are examples of visual cues •• Always look for functionalities.

Fun Fact: All of the GUI elements you see on a computer screen began on some graphic designer's drawing board.

GIGO

GIGO = "Garbage in,
Garbage out"

noun Computers.

a rule of thumb
stating that when
faulty data are fed into a
computer, the information
that emerges will also be
faulty.

Computers are all as dumb as a box of rocks and/or a bag of hammers (pick your analogy).

Indeed computers only know what we tell them.

Therefore we must be very careful what we tell them (via *mouse clicks*, *taps* & *gestures* and *keyboard presses*).

Clicking the computer mouse willy-nilly, full-tilt boogie, college party animal (that is **without** thought and intention) is a contemporary example of GIGO.

1981 flashback

During the autumn of 1981, your author wasted a lot of punch cards trying to get a mainframe computer to print his name (which was his first assignment for his high school "Introduction to Computer Science" class).

He's learned a lot since then during his now 25+ years as a Public Librarian / Tech Wizard helping people with their computers and various & sundry gadgets — hence this book.

From GIGO to "Computer Comfort"

Introducing (Or Reintroducing) the Computer Basics

Over the next few chapters, we will quickly and casually cover the computer basics which are few in number and easy peasy lemon squeezy to learn and to perform.

They're standing by, waiting in the wings to help you get stuff done ✔

Due to space considerations (and to prevent your eyes from glazing over) we will be painting the computer basics in very broad strokes so to speak.

If you would like some **additional instruction** on the computer basics, we have you covered with a *plethora* of how-to videos 🎞 Please see the "Where to Go From Here" chapter for details.

The mouse

The mouse (regardless of its physical appearance and how many buttons it has and if it has a scroll wheel or not) can be used for *selection*, *input*, *navigation*, and *scrolling*.

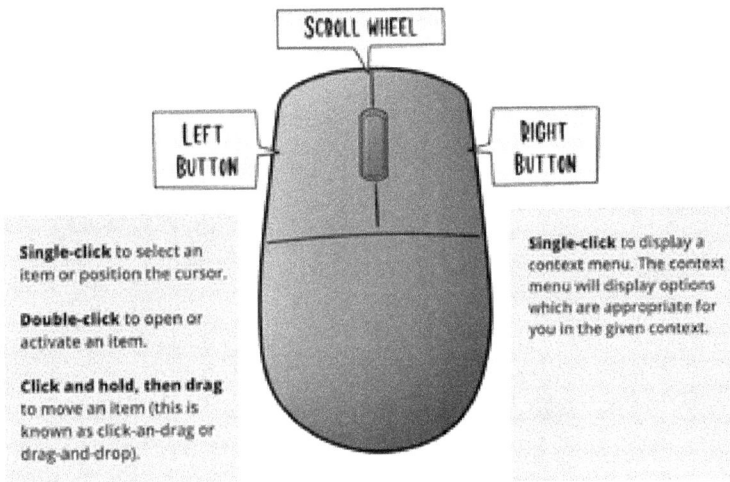

SCROLL WHEEL

LEFT BUTTON

RIGHT BUTTON

Single-click to select an item or position the cursor.

Double-click to open or activate an item.

Click and hold, then drag to move an item (this is known as click-an-drag or drag-and-drop).

Single-click to display a context menu. The context menu will display options which are appropriate for you in the given context.

The mouse is a simple but very powerful tool to help you get stuff done ✔

Moving focus

As you may already know, you can move the mouse with your hand to move the *screen arrow* around your computer screen:

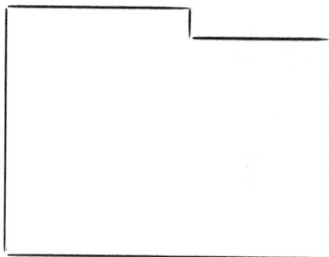

Photos '21

Here we see a screen arrow en route to a specific destination.

But it's important to keep in mind that the exact position of the screen arrow will affect the outcome of a mouse click.

Actions often have reactions.

Stay tuned for context menus!

Selection

You can click on a file **one time** to select it (which will put a visual cue "highlight" effect around the file):

FILE NOT SELECTED

FILE SELECTED

Input

Using the mouse to put the *blinking line* (i.e. text insertion point) into a text box is a type of input:

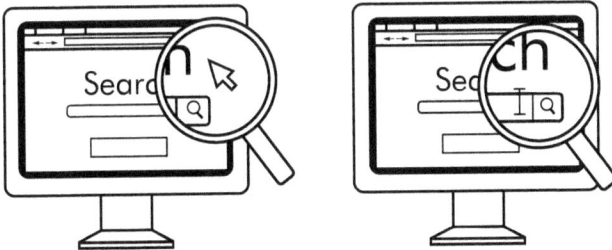

Visual Cue: Here we see a screen arrow becoming an "I-Beam" when it is placed over a text box.

Making a menu selection is a type of input as you are giving the application a **command** (e.g. "Save" document):

Abstract application "File" menu

State

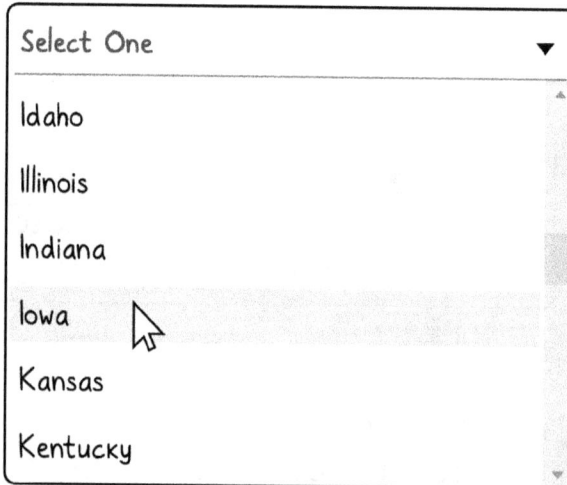

Abstract web form menu selection

And using the mouse to "hover" over something on your computer screen to cause a *tooltip* to appear is a subtle form of input:

Before screen arrow "hover"

Contains the files and folders that you have deleted.

After screen arrow "hover"

Navigation

Brew a nice cup of tea and use the mouse to "noodle around" your computer to get familiar with its filing system:

Abstract "Pictures" folder

You can use your computer's *search box* to quickly search for files by *file name, file type,* and words that you know are in the file that you're looking for. That being said, it pays to be *digitally organized* (e.g. pictures go in the "Pictures" folder and documents go in the "Documents" folder).

Using the mouse to cause a *context menu* to appear (more on those guys later) is another form of input and it's worth repeating here that actions often have reactions:

Scrolling

There may be more to see! You'll need to look "below the fold" to find out!

Forms and web pages will often be longer than what can be shown on a computer screen or window — which necessitates *scrolling*

Visual cues for scrolling include arrows and bars.

They can be on windows...

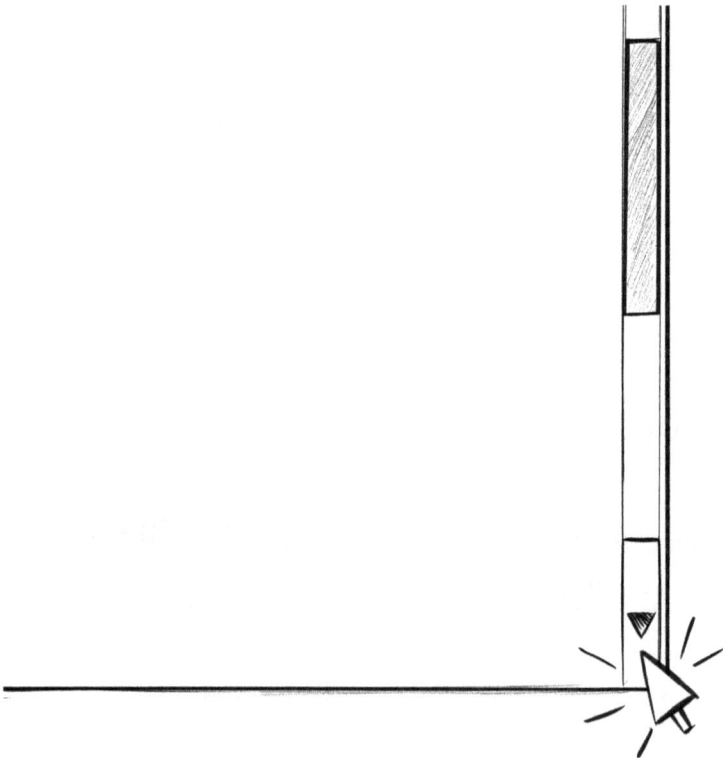

And on menus...

State

Select One	▼
Idaho	
Illinois	
Indiana	
Iowa	
Kansas	
Kentucky	

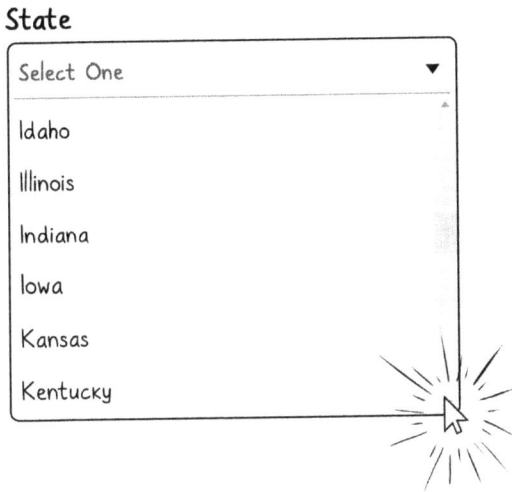

You can also use the mouse's scroll wheel (if it has one) but you must **first** click inside whatever you wish to scroll (such as a long web page or computer window) to put *focus* there.

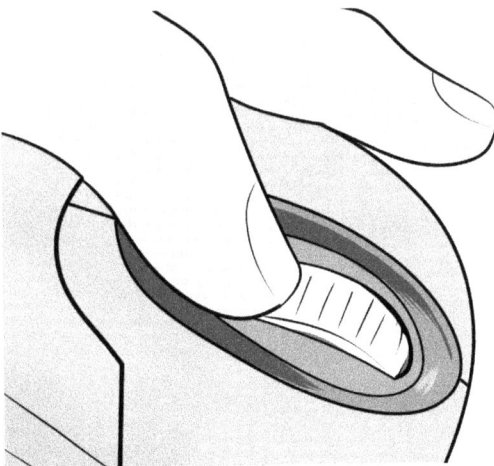

Don't go full-steam potato

Many people have made the mistake of using the mouse as if they were playing a *pinball machine* (that is without thought and intention) and wind up getting lost in the proverbial weeds as a result.

Context Matters

Context matters as much in the computer world as it does in the real world. For example, you would (probably) not want to sing your favorite aria from La Bohème while standing in line at the supermarket.

Some things are better left in the shower…

(Especially if you can't sing 🙀)

Some examples of computer context...

Context includes which window or folder or directory you currently have open (if any)...

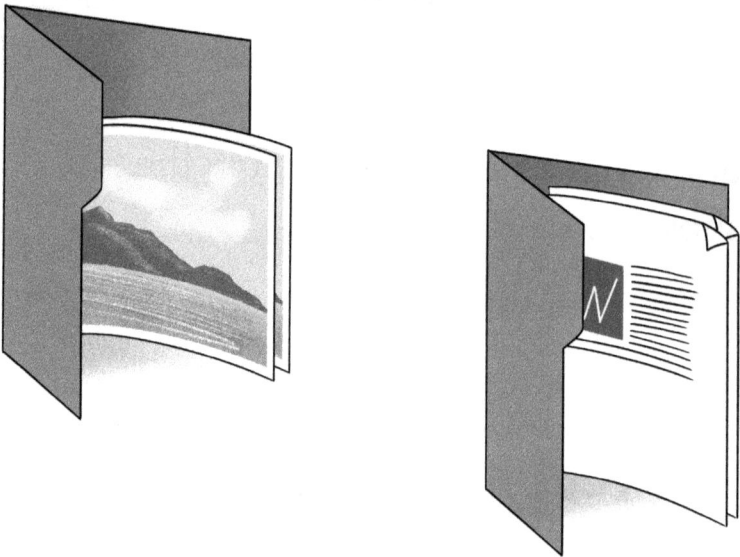

Above we see abstract versions of a "Pictures" folder and a "Documents" folder. (We use abstract illustrations for examples because appearances will vary between computers but visual cues will always be present 👀)

Context includes your computer
desktop…

_____

The *computer desktop* is so
named because it is very
similar (at least
conceptually) to the top of a
real-world desk.

Context also includes what the tip of
your screen arrow is currently over...

"Bea" inadvertently closed her web
browser window through what was
frankly a careless mouse click while she
was in the middle of completing an online
job application and subsequently lost
everything 🐾

And context includes whichever app
you are currently using…

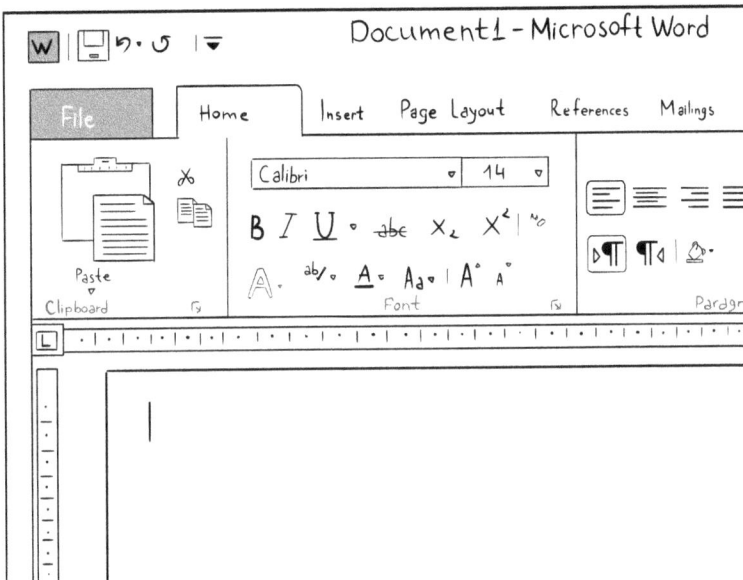

NOTE: There is a *plethora* of visual cues
in the above illustration including app
name, file name and various icons for
functionalities (e.g. cut ✂, paste and
clipboard 📋).

Key points

1. Context will affect the outcome of mouse clicks, keyboard presses and taps & gestures (on *trackpads* and *touchscreens*).

2. Know which app you are currently using (if any).

3. Know exactly what the tip of your screen arrow is over before you click the mouse or tap your trackpad.

Did I mention context matters?

44

About the Double Mouse Click

How to double-click

1. Hold the mouse very still.

2. Press down and quickly release the left mouse button two times in quick succession ("knock knock").

"To be successful in life, you must get in the habit of turning negatives into positives." -- George Foreman

If you have trouble double-clicking the mouse...

You can single-click a file icon or an application icon to select it and **then** press the enter / return key on your computer keyboard.

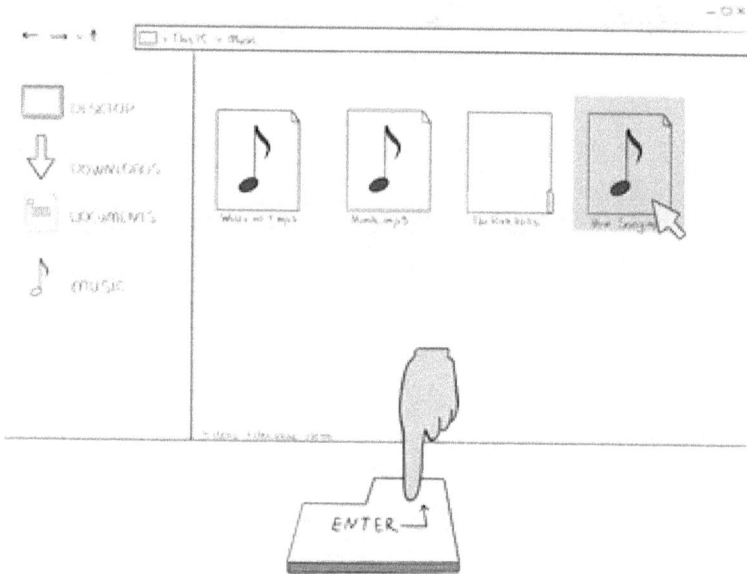

Question: When do you have to double-click the mouse?

If you wish, you can double-click icons on your computer desktop and things that are inside a *computer window*:

If this doesn't make sense 🤔, single-click everything.

If nothing happens after the single mouse click, you can press the enter key / return key on your computer keyboard or try a double-click.

The "Secret Agent"
Technology Tool

"Context menus" (which are also called "secondary menus" or "pop-up menus") are like secret agents because they remain hidden until summoned.

The selections on a context menu will vary depending on…

You guessed it! **Context**!

When it comes to context menus, context will be determined by where the tip of your *screen arrow* is; that is what it is currently over.

Context menu example

How to summon a context menu

You can summon a context menu through a *secondary mouse click*.

Exactly **how** you can secondary mouse click will depend on which computer you are using and which *mouse* you are using.

But generally, you can click the right mouse button to summon a context menu:

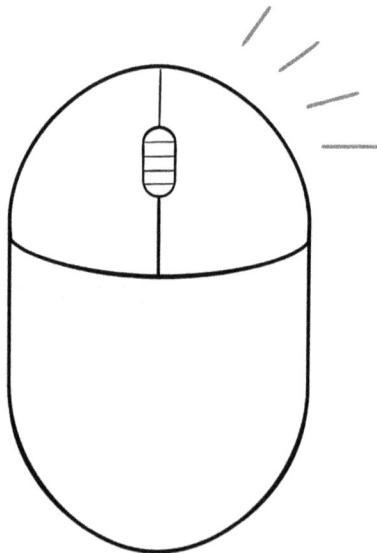

If your mouse has only one mouse button — you can try holding down the control key (or the alt key) on your keyboard as you click the mouse…

You can also *search the web* for how to secondary mouse click on your computer.

The context menu is a very powerful tool. Try the secondary mouse click on various things on your computer screen and peruse the context menus that will pop up.

Breaking Down the Computer Keyboard

The original QWERTY layout was developed by Christopher Latham Sholes. He filed a patent application in 1867 for an early version of a typewriter.

The letters and numbers keys on a computer keyboard are identical to a typewriter.

FMI
visit
www.technologyshy.com

Special function keys

Look over the computer-specific keys (i.e. non-letters and numbers) on your computer keyboard.

In addition to the time-honored QWERTY keys, there may be special function keys near the top of the keyboard above the numbers / symbol keys.

These special function keys could control things like screen brightness ☀ or your computer's sound volume 🔊 or summon a *Help Menu* specific to the application you are currently using — the possibilities are endless!

Depending on your computer, these special function keys could literally change depending on which application you are currently using.

By all means, press these special function keys and see what they do.

But please keep in mind that the **outcome** may depend on *context* — that is which application you are currently using (if any) when you are pressing a special function key.

Once again, use casual experimentation for the modest but very important goal of *familiarity*.

Some special function keys will be more intuitive than others.

Keyboard shortcuts

You can use keyboard shortcuts to give the mouse a well-deserved rest $_z$z^z

The keyboard shortcuts you can use will **vary** depending on which computer you are using (or more accurately the *operating system* that is on the computer you are using) and the keyboard that you are using with your computer.

Look at the keys on either side of the spacebar on your keyboard:

These are the keys you can use to bring about time-saving (and wrist-saving) keyboard shortcuts such as this one:

Procedure

We will use the "undo" keyboard shortcut as an example here.

1. Hold down the "ctrl" key (on a Windows keyboard).
2. Press the "z" key on the keyboard.
3. Release the "ctrl" key.

You can *search the web* for your computer's keyboard shortcuts (e.g. "Windows 11 keyboard shortcuts").

Like everything else related to computer use, keyboard shortcuts require **patience**, **persistence** and **practice** (and how-to videos for them are available on the Help for the Technology Shy website).

The specific key combinations of the keyboard shortcuts will depend on which computer you are using.

For example, the "undo" keyboard shortcut (very handy if you make a mistake) will be **control** - z on a *Windows* computer and **command** - z on a *macOS* computer.

Many people have caused unexpected (and counterproductive) things to happen due to the fact that they **inadvertently** performed a keyboard shortcut or pressed a special function key.

Quick Tip: If something unexpected happens, try pressing the "escape" key on your keyboard.

Let's Go FORMal

To be sure, forms are not the sexiest of computer-related topics

But a form of some kind will figure into pretty much everything that you can do with a computer.

You have to use a form (or a type of form) to...

- Address, compose & send email messages
- Search the web
- Shop online
- Use virtually any online service (e.g. medical portal, banking, insurance, library e-books, et al.)
- Apply for a job

As with everything else related to computer use, you need to be visual and very attentive when filling out a form — just as we do when filling out real world paper forms.

Be sure to look at each form element **individually** (e.g. text boxes, checkboxes, radio buttons, and menus) and understand what information it requires before you interact with it.

Read any instructions very carefully.

Apply the same focus & methodical procedure you would use to complete a paper form to completing a form on a computer.

Search

WHAT IS A EXE FILE?

GO

Fun Fact: When you search the web, you are using a form.

Fun Fact: You can often press the "enter" or "return" key on your computer keyboard to submit a form.

Address

Address Line 1

City

State

Select one ▼

Postal Code

Menu boxes are common in forms, look for visual cues (like the one circled above).

Cats
Only you have access

Share this folder

Only people invited: can edit ⌄   Settings

Henry

𝒢 Create then copy link   Share folder

Forms are not usually this cute but they often provide useful functionalities.

Mainframe Computer

The GIGO principle also applies to forms.

The Power of Selection

Selection and *context* are important factors during computer use.

For example, here is a selected picture file being **copied** from the *"Documents folder"* to the *"Pictures folder"*:

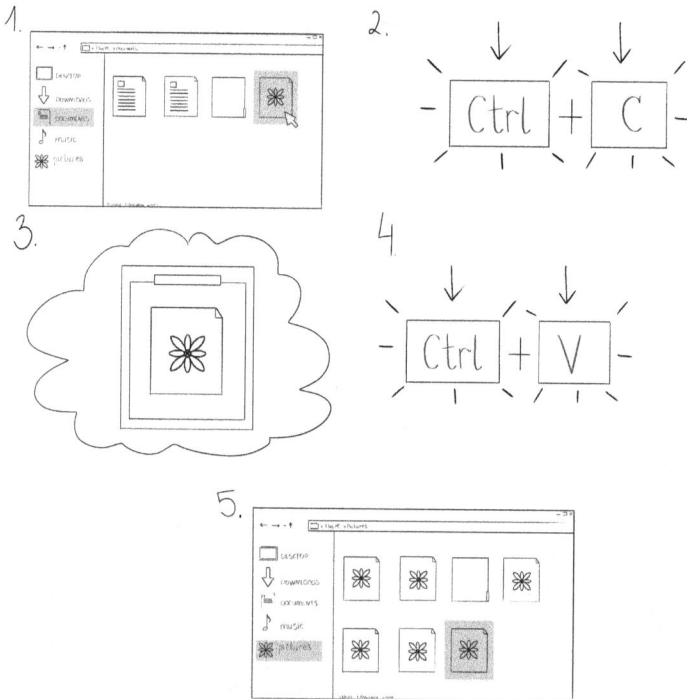

Please note the visual cues and the "invisible clipboard" (#3) that will temporarily hold copied items.

The "copy" keyboard shortcut (or menu selection) would copy **only** the selected text in this document:

You could then *paste* the copied text to another part of the document and/or into another file via the paste keyboard shortcut or menu selection.

It is much easier to **show** the various ways of copying & pasting text than to attempt to describe them. This being the case, visit www.technologyshy.com/selection.

Just Noodle Around

"The oldest and strongest emotion of mankind is fear, and the oldest and strongest kind of fear is fear of the unknown." - H. P. Lovecraft.

Noodling around a computer — and apps on that computer — will lead to **familiarity**.

And getting familiar with a particular computer or app will increase your **confidence** with it.

And with increased confidence comes increased **proficiency**.

This being the case, brew a nice cup of tea and give your mouse a workout.

Use the mouse to click various things on your computer screen just to see what happens.

Your only very modest goal here is familiarity.

Search for It

Got a question? Is there something you need to find out? You can *search the web.*

MIKee

What's the difference between marinara & spaghetti sauce?

Q All   ⊘ Shopping   🖾 Images   ⊚ Maps   🖃 News   ⦂ More

Marinara is tomato sauce, but it's a thinner, simpler sauce that cooks very quickly: It only needs to simmer for about an hour.
Spaghetti sauce is a version of marinara, but it usuarlly contains additionals ingredients, like meat or vegetable.

Marinara vs Tomato Sauce: What's the Difference?

1. Sit down in front of a computer that is *connected to the Internet.*

2. *Noodle around* for a web browser application.

3. Type what you are looking for into the web browser's *search box:*

     ⌕ help for the technology shy

4. Press the *Enter / Return* key on the keyboard or click the web search box's *"Go"* button (if it has one).

5. Peruse the *search results page*.

6. Follow visual cues (e.g. text that are actually *hyperlinks* and *navigational breadcrumbs*) to visit web pages and to return to your *search results page*.

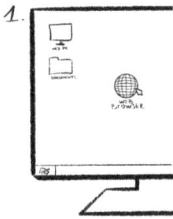

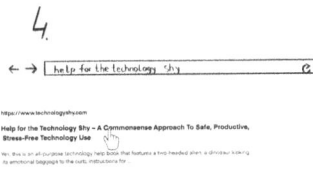

If you are sure of the *web address* of the website that you wish to visit, you can type that in instead.

www. technologyshy.com

Search your computer

1. Look around your computer screen for a search box or search icon 🔍

2. Type in what you are looking for (e.g. names of files, folders, applications).

3. You can usually *double-click* a file or application to open it.

gourmet cat treats| 🔍

Positive ID

Knowing — and recording into your technology notebook — what you are using (e.g. your computer's brand and operating system) will help you…

1. take control of your computer use

2. find out which *peripherals* you can physically connect to your computer (if you need to do so)

3. get *good help*

Procedure

Brew a nice cup of tea and take a good look at your computer; look for logos and brand names.

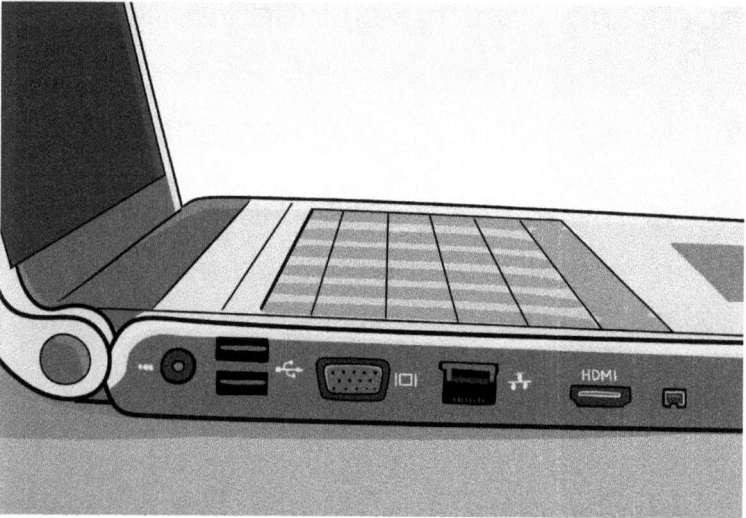

You may find it useful to know which ports are on your computer as they enable you to physically connect peripherals to your computer such as printers, speakers and external storage devices.

Searching for "about" (not as existential as it sounds 🫤)

A computer's outer shell will usually reveal its brand (e.g. Dell, HP, Apple) but to find out which *operating system* is on your computer (e.g. Windows 11, macOS 12), you're going to need to locate its "About" screen.

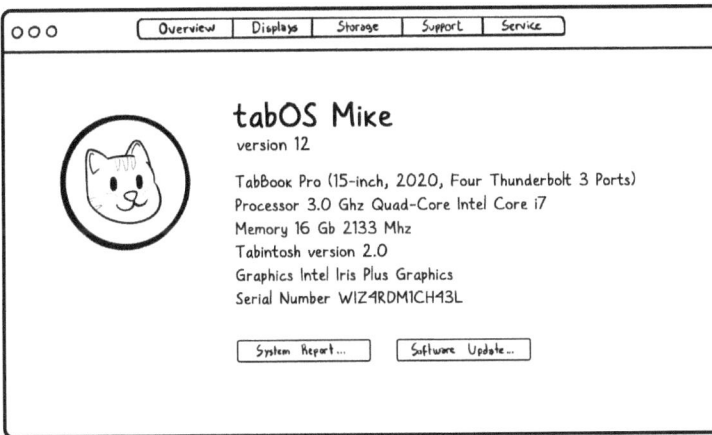

Abstract illustration of an "About" screen

Sip your tea and use the mouse to *noodle around* your computer for its "About" info.

You can also try cutting to the chase as it were and type "about" into your computer's *search box*:

Abstract computer search box

Careful documentation will never go the way of the Dodo.

Record your computer's brand, operating system and *model number* and/or *serial number* into your technology notebook.

There's an App for That (Probably)

A computer could have state of the art *hardware* and the latest *operating system* but it would still be a high-tech paperweight without *programs* or *applications* (or "apps" as they are commonly called).

Applications enable us to **do things** on computers such as…

• create & format documents, spreadsheets and presentations

• send and receive messages

• check the weather ☔

• create calendar events and reminders

• and of course, *search the web* (for anything you need to find out)

Fun Fact

When you use an app to create and save something to your computer (e.g. a document), you are creating a *file*.

Recap: During your casual and low-key *noodling around* sessions, you used the mouse to open and close various *applications* on your computer just to get a sense of them; just to kick their tires so to speak.

What does it do?

If an app's name does not indicate its functionality (e.g. "calculator", "calendar" & "reminders"), assessing the app's *graphical user interface* may shed some light on its purpose.

You can *search the web* for what an app does.

And of course, you can simply noodle around with an app (with no pressure, urgency or expectations) by using the mouse to click anything you see on the app's *graphical user interface* such as buttons and menus.

And we see plenty of those things in this illustration…

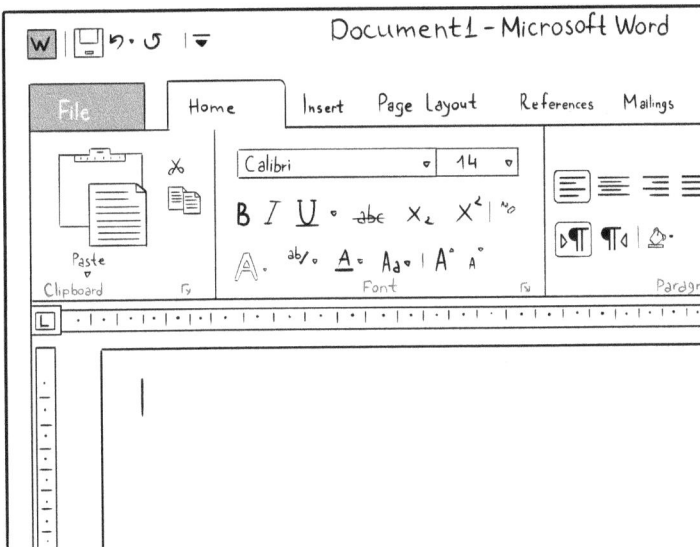

It's best to noodle around with an app first <u>before</u> you actually need to get something done with it ✔

Any questions about a particular app?

If so, write them down in your technology notebook for later review and clarification.

If you are not sure... 🤨

... which app you need to use for a particular task on your particular computer, you can ask your employer (if applicable) and/or *search web*:

Search

How to edit a photo on a macOS computer?

GO

It really helps to know which operating system is on your computer.

The right tool for the job

A will or testament is a legal document that expresses a person's wishes as to how their property is to be distributed after their death and as to which person is to manage the property until its final distribution.

The Setup: An eighty-something clergyman with a delightful sense of humor was using one of the library's public computers to create his will.

The Surprise: "Theodore" used a spreadsheet application (e.g. Microsoft Excel) instead of a word processing application (e.g. Microsoft Word). He had typed his entire will into a single cell of the spreadsheet document → and then he wanted to format it 😶

The Denouement: In a nutshell, Theodore used the wrong app. As in the real world, just because you **can** do something (like singing your favorite aria from La Bohème in the supermarket) doesn't always mean you **should**.

The Workflow Workaround: We thoughtfully used the mouse to *copy and paste* Theodore's will from the spreadsheet cell into a Microsoft Word document → where it could then be easily formatted.

The Keys to Theodore's Success:

1. a clear *objective* 🤷
2. getting *good help* (which was yours truly ☺)
3. patience; a willingness to methodically work through the technology-related nuts & bolts 🔩
4. acceptance 🕊 (Theodore was very calm during the entire procedure)
5. a willingness to do what needed to be done ✔

Files

Files are things you can create with *apps*.

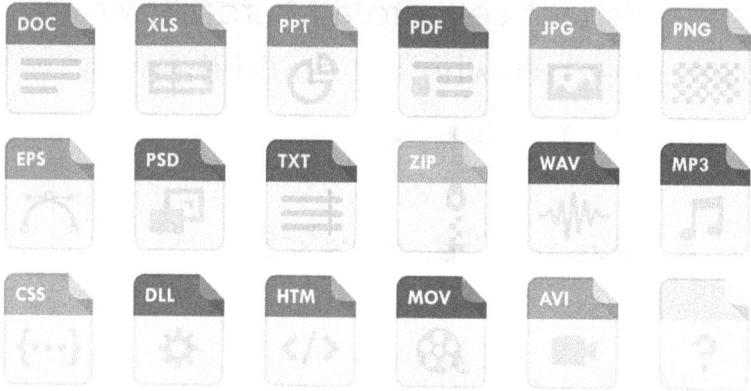

You can search the web for file name extensions (e.g. .mov) to learn about that file type.

You could, for example, use a word processing **application** (e.g. Microsoft Word) to create a document **file** (e.g. .doc).

There are also *executable files* which can install new applications onto your computer.

Executable files can be benign or malevolent.

You could *methodically* download an executable file and use it to install a useful application onto your computer.

But on the other hand, my technology shy friend Brandon **inadvertently** opened an executable file and subsequently installed *malware* onto his computer.

Some rules(s) of thumb for staying out of trouble:

1. When it comes to downloading new applications to your computer, it's best to stay within your computer's *App Store* (which you can find on your computer by noodling around).

2. If you are not sure what a file is, do not do anything with it.

3. Be especially wary of *installation windows* → it pays to be visual 👀

4. Files attached to email messages can also be benign or malevolent.

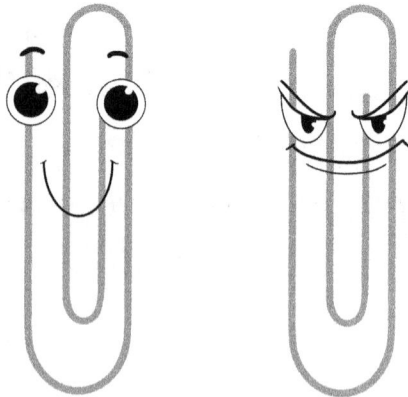

Start From Your Drawing Board

When it's time to *get stuff done*, it is important to have a clear idea of what you wish to accomplish before you go anywhere near a computer.

You may wish to record your overall
goals at this particular moment in time
on a single piece of paper or napkin...

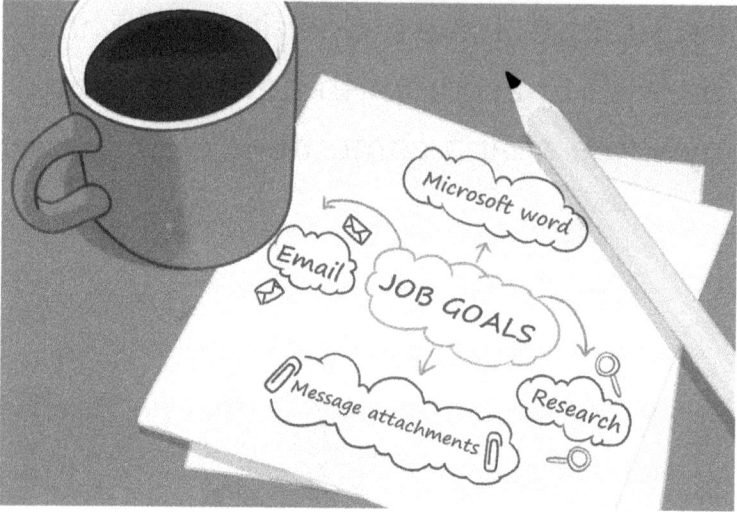

About multitasking...

*The short answer to whether
people can really multitask
is no. Multitasking is a
myth. The human brain cannot
perform two tasks that
require high-level brain
function at once.*

Break each of your objectives into more manageable pieces...

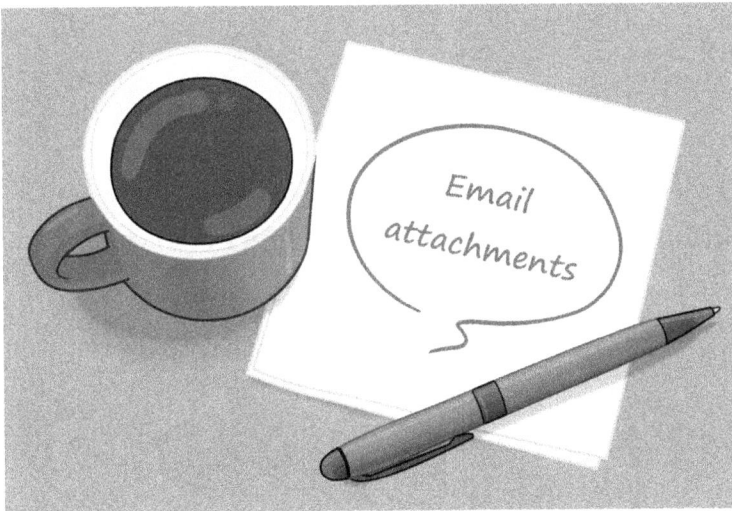

This will help you focus on one particular goal (i.e. task at hand) as you are using your computer.

REMINIDER: If you are uncertain 😐 about anything you need to do, get some *good help*.

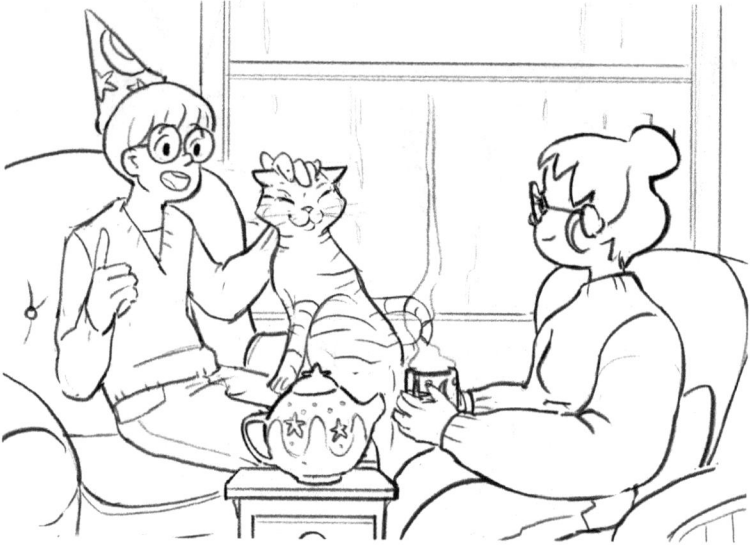

"You can access your email through your email provider's website and through an app on your computer."

Think About What You Want to Do 🤔 & Then Look Around for a Way to Do It 👀

Check out the functionality-revealing icons and text on this abstract app interface as well as the "Settings" wheel at the upper right corner.

Preparation Recap

- **after** you have familiarity with the computer from *noodling around* with it...

- **after** you have familiarity w/ the apps you need to use...

- **after** you have an objective firmly in mind 🙌 or better yet written down or sketched out ✏️

- you can **then** brew a nice cup of tea and sit down in front of your computer in full *getting stuff done* mode.

General procedure for using an app

Quite often, the *"step-by-step instructions"* for anything you wish to do will be right in front of you (as long as you are using a suitable app) in the form of visual cues, which include buttons, icons & menus.

Every time I host a tech meetup, there's always one person who is scribbling furiously <u>instead</u> of focusing on understanding how a particular app works (which would be much more beneficial to them).

```
Methodical meaning
     (of a person)
orderly or systematic
in thought or
behavior. she was so
methodical, she kept
everything documented"
```

vis·u·al meaning
adjective:
relating to seeing or
sight. "visual
perception"
noun:
a picture, piece of film, or
display used to illustrate or
accompany something.
the music should fit the
visuals"

Real world correlation

What do you do when you are driving
through an unfamiliar neighborhood?

You would not (one would assume) push
the accelerator to the floorboard and
make random turns. The odds of you
arriving safely at your destination would
be astronomical at best.

You would instead proceed at a measured pace while looking for visual cues (e.g. street signs) en route to your destination.

Sightseeing is the real world parallel to *noodling around* a computer.

Whenever you are confronted with an unfamiliar interface (or an unfamiliar **anything** for that matter):

1. Sip your tea ☕
2. Breathe.
3. Look around; take in what you see…
4. Refer to your objective, your current task at hand.
5. <u>Before</u> you click or tap anything, mentally break the interface down, piece-by-piece, section-by-section…

6. Look for what you can use to accomplish your current task at hand (i.e. functionality).

7. Look for visual cues (e.g. back and forward navigation buttons).

8. Work methodically toward your objective (just as you would safely cross a pedestrian intersection).

before you click or tap

For example, if your current task at hand is to search your computer, you would logically **not** want to disable its search functionality:

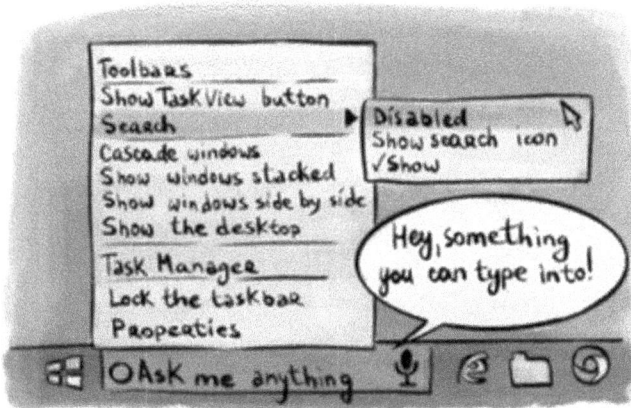

Note the position of the screen arrow before you click the mouse.

97

Under your command, the computer mouse and keyboard can be drunken layabouts **or** diligent workers.

And just as a reminder…

Context Matters

Don't Be Sisyphus

In the Greek myth of *Sisyphus*, the title character was a king cursed by Zeus to forever roll a boulder up a hill in the depths of Hades, only to always have it roll back down.

Don't be that guy!

Project completed! Good job!

Why ask why?

Your Tech Wizard Tour Guide has worked with many people over the years who simply would not let themselves be **done** with a computer-related project.

For example, they would agonize over why a particular *application* or *online service* did not work as expected.

If their project had been completed... **If** they had accomplished what they set out to do, I would simply state that sometimes *things are what they are*.

In a nutshell: When your clearly defined objective has been realized, when you have hit your target, when you have reached your goal, stick a fork in your project and then move on.

"Viruses"

Once again, we're going to paint in very broad strokes here 🖌 Because really from a practical standpoint, there's no point in doing a "deep dive" on viruses.

(Remember the Help with the Technology Shy mantra *Keep It Simple, Snookums*).

When people speak of computer *viruses*, they are usually referring to *malware*.

Malware is intrusive software that is designed to damage and destroy computers and computer systems. Malware is **a contraction for "malicious software."**

Malware can seriously degrade the performance of your computer and lead to the theft of your personal information so it's best to nip it in the bud → pronto.

Examples of common malware include *viruses*, *spyware*, *adware* and *ransomware*.

The specific permutations of *malware* are seemingly endless and it would be pointless (again from a practical standpoint) to try to cover them here. Our time would be better spent counting all of the stars in the sky ✦

It's important to note that the signs of malware infection can be **ambiguous**.

For example, a computer might be performing very slowly due to malware infection but also perhaps due to its age, lack of *maintenance* or the presence of extraneous, unnecessary, annoying, resource-hogging *bloatware* 🌚

At the time of this writing, "bloatware" often comes **preinstalled** on computers running the Microsoft Windows operating system.

Such *software* is **not** created by Microsoft and is also, by the way, superfluous and counterproductive.

Want a plan? Here's a plan 📋

If your *web searches* are being redirected to ads, stop using your computer and get some *good help* — which we will discuss how to do in the next chapter.

There will be advertisements on web search pages but if your *search results pages* look "tacky" then your *web browser* could be infected with *adware*. Granted, the term "tacky" is subjective, so you can *search the web* on other computers (e.g. a public library computer or a friend's computer) and compare the *web search results pages* on those computers to your own.

If you suspect your computer has adware then stop using it and get some *good help*.

If you get a pop-up message informing you that your computer is "at risk" then stop using your computer and get some *good help*.

If you feel threatened or coerced in any way then stop using your computer and get some *good help*.

Do not provide any confidential information upon unexpected request or demand from any source including email messages 🤐

And please don't fall for any scams, swindling, shenanigans or skullduggery 👿

Use common sense 🤔 If something seems too good to be true, it most likely will be.

A computer use cautionary tale

An agreeable chap whom we will call "Dennis" outsourced the administration of his computer to a shady organization resulting in malware infection and a host of other bad things.

We all need to take <u>full responsibility</u> for our technology use — it can be dangerous to try to avoid it.

Factory reset

If you suspect your computer has been infected with malware, the quickest and surest and safest solution is update your *technology notebook*, *back up your files* and then *factory reset* your computer.

A factory reset can be a fresh start and a clean slate for your computer (and your computer use).

Remember to carefully follow instructions and always look for visual cues.

How to factory reset your computer

If you know *which computer you are using*, you can begin by *searching the web* and/or contacting your computer's customer support department.

From there, it will simply be a matter of proceeding methodically; following visual cues and reading the instructions on each screen very carefully — and getting *good help* if you need it.

Factory resetting a computer is also the subject of a Help for the Technology Shy blog post 🖊 and a how-to video 🍿

Getting Good Help

"Kim" came to the library with a smartphone which at first charged very slowly when plugged in and eventually would not charge at all.

After researching this issue and following some non-intrusive troubleshooting methods (e.g. *restarting the smartphone*), it began to at least appear that this was a mechanical issue.

Suddenly, a well-meaning jackass appeared and said "May I jump in?"

My first and immediate inclination was to politely but firmly say "**No, thank you**" (because I knew what was coming).

Unfortunately the jackass was too quick for me 😣

The jackass happily told Kim she could clean out her smartphone's charging port with a paperclip 📎

Please don't do this!

Intuitiveness and good judgement should tell us that sticking a pointed metal object into a port filled with delicate electronics would not be a good idea 🤔

You can often identity a well-meaning jackass by their giddiness and lack of thoughtfulness.

Well-meaning jackasses also tend to oversimplify technology-related projects which may require multiple apps, online services, additional hardware, et al.

Other people to look out for are *know-it-alls* — these are smug, arrogant usually charmless people whose primary and perhaps only goal is to project their self-importance and a pox on them.

And there are also *nogoodniks* who would love to scam you, sell you something you do not need and/or literally take your computer hostage through the installation of *malicious software* → and a pox on them too!

Good eggs

For our purposes, *good eggs* are people who genuinely want to help and have no ulterior motives.

Good eggs are calm, attentive, mature, thoughtful and objective (i.e. they are not trying to sell you something and have no ego to boost and/or a particular technology-related ax to grind).

Rule of Thumb: Seek out calm, positive, mutually respectful collaborations.

Spilt milk 🥛

(It was actually coffee creamer.)

During the production of this chapter, a nearly full container of coffee creamer was spilled onto the keyboard of my notebook computer.

At this very moment in time, my notebook is an expensive high-tech paperweight. This is "major bummage" as my friend Sandy would say but hey, we live in the material world and these things happen.

Thankfully, since this very document you are now reading was *cloud synced* between my notebook computer and my tablet, I was able to resume working without having lost so much as a word.

I mentioned this spilt milk mishap to my neighbor and he said he had a similar happenstance and helpfully related his very positive experience with a local "computer nerds" company 🤓

With this non-biased endorsement in mind, I wrote down the nerds' company name and nearby location into my *technology notebook*.

I also called my computer's tech support number (one of the many benefits to *positively identifying your computer*) but alas they were unable to help me revive my dead-as-a-door-nail notebook computer over the phone.

They transferred me to customer support and they in turn gave me a short list of local authorized service providers.

If you don't know which phone number to call for your computer's **actual** technical and customer support services, you can borrow a friend's internet-connected device and carry out a web search or ask a librarian.

VISIT YOUR
LIBRARY

How to deal with any computer-related challenge (the short version)

- Make a nice cup of tea

- Breathe.

- Be able to clearly state your objective or issue (e.g. "My notebook computer will not power on after creamer was spilled onto its keyboard").

- Document the issue in your technology notebook.

- Get *good help*.

Yikes! A strange or alarming message appears...

1. As always, stop, look and think before you click the mouse or press any keys on your keyboard.

2. Understand the *context* – for example, is this message on your *computer desktop* or is it on a web page?

3. If the message is coming from a web page, it will be a bluff → quit your web browser and *clear your web browser's cache* if the threat persists.

4. If the disconcerting message pops up on your *computer desktop*, analyze it. Is the message trying to scare you or does it come across like a statement of fact?

5. If a message directs you to call a phone number and/or provides a link to a website, it will be a scam.

Legitimate system messages are those that are actually coming from your computer's operating system and you need to be able to recognize them.

Legitimate, matter of fact system messages look something like this.

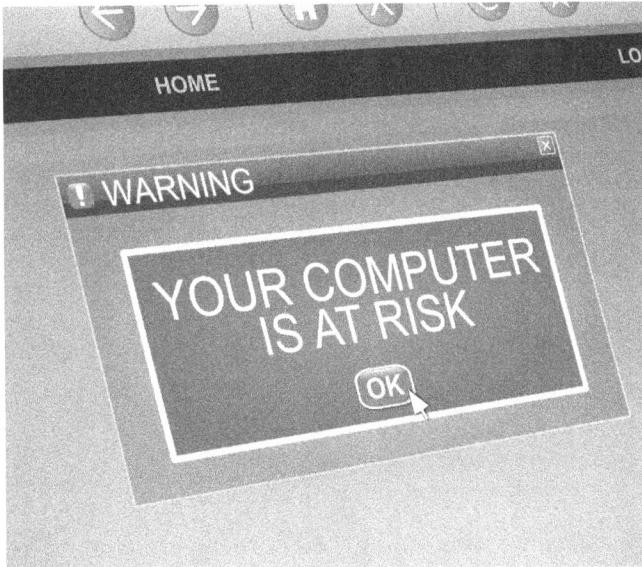

It's <u>not</u> OK! Please do not engage with messages like this.

Suggestion

If threatening messages are emanating from your computer itself, it is most likely due to *malware infection*, and my advice is to *back up your files* and *factory reset* your computer.

If you are getting *bloatware* messages, carefully write down the information on the bloatware message and then *search the web* for *how to remove it from your computer*.

A pop-up message offering you a 30-day free trial of whatever is most likely going to be bloatware.

Sources of good help

1. Good eggs (including of course your stalwart and intrepid Tech Wizard Tour Guide)

2. Authorized service providers

3. Online support communities / forums

Online support communities

There are good eggs galore in online support communities.

Support communities may be available for your particular computer or operating system or mobile device or online service that you are using.

Online support communities are (for the most part) comprised of people who are simply users of various computers, operating systems and online services → and not employees of the company.

(Although company technologists may moderate the support forums.)

The knowledge and skill levels of the people you will "meet" in these online support communities will vary between beginning, intermediate and advanced.

Keep in mind that some people will also be promoting their own services.

This does not mean that the information they provide will be false. That being said, a good policy is to back away from arrogant people. And don't buy anything.

If possible, **cross-reference** what you are told by other support community members (preferably those folks who simply want to help).

Support forum procedure

1. Search the web for the support community for your computer's operating system or the online service you wish to use.

2. *Proceed methodically through the support forum* to post your question(s) and/or search for answers.

3. Write down any potential answers or useful information into your technology notebook.

4. *Check your email* for answers to any questions that you posted.

"Where's My Stuff?"

The most often asked question on the Public Library Tech Help Front is "Where's my stuff?" People cannot find the *files* they created.

An engaging and optimistic artist we will call "Olga" had borrowed one of the library's public laptops and used it to create a manuscript about her father who was a notable Russian historian.

Olga knew which *app* on the laptop she could use to create and format her manuscript.

(So far so good 😊)

Olga created a file and gave it a *descriptive file name* 👍

But…

(And here's where Olga took her eye off the ball ⚾)

Olga had saved her manuscript file to the "Documents" folder on the Library's laptop instead of onto her *external storage device* (which was a flash drive).

Please note the *visual cues* in this *Save as window* – Including **This PC > Documents** (which I call a "navigational breadcrumb") in the location box:

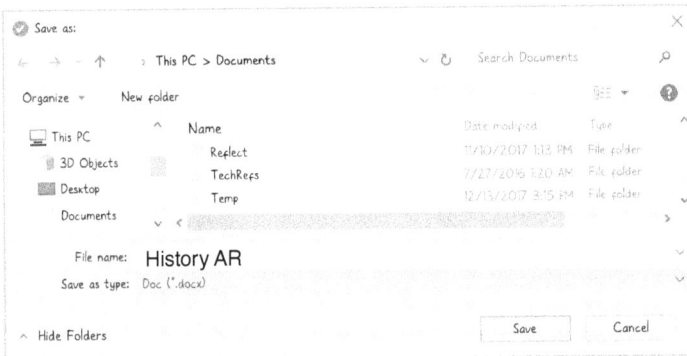

Organizational procedures and mandates come into play.

On a Library laptop, whenever someone stopped using it and signed out of it, the laptop would automatically be wiped — all of the files that had been saved onto the laptop would no longer be there the next time anybody signed into it.

And this is what had happened with Olga.

1997 flashback

Your Tech Wizard Tour Guide inadvertently saved his Master's thesis onto the University's computer network when he **thought** he had been saving it onto a floppy disk 💾

As a result, the network went down and your author's gosh darn awful thesis on the films of Alfred Hitchcock went with it — POOF!!!

The moral of these stories...

Be visual 👀 Know **exactly** where you are saving your work → especially if you are using a public computer.

The Pomodoro Technique

Staying focused during computer use is
<u>essential</u>. Try this…

THE POMODORO TECHNIQUE

Step 1
Choose the task

Step 2
Set a timer for
25 minutes

Step 3
Work on the task
until timer beeps

Step 4
Take a short break
of 3-5 min

Step 5
Repeat the cycle 4 times take
a longer break after 4 sessions

Visit your local library and ask for books
and other materials on how to stay
focused.

A Quick Visual Reminder 👀

Things like this happen <u>a lot</u> during computer use and they are easy to overlook.

For example, clicking a *link* on a web page can open the web page that the link leads to in a *new tab* within the same web browser window:

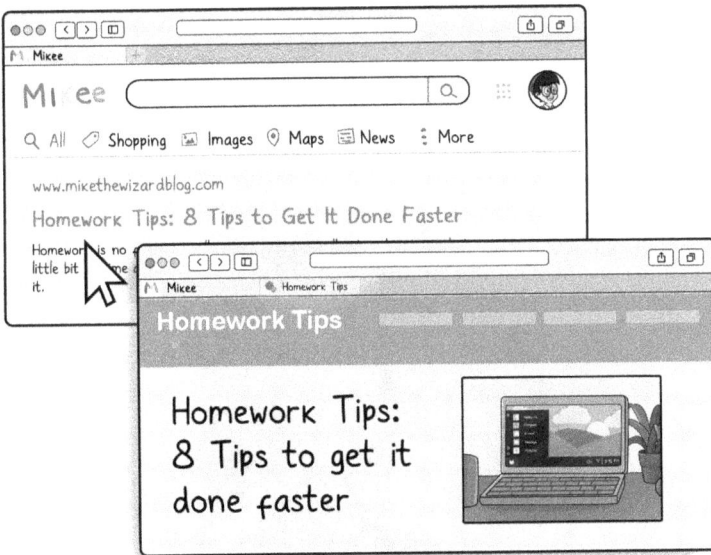

Always be observant during computer use and expect the unexpected.

Troubleshooting Made <u>Really</u> Simple

If something perplexing and/or unexpected happens, your Tech Wizard Tour Guide Grease Monkey has a very simple formula for you.

You won't have to get your hands this dirty.

1. Document the issue.

2. Stop using your computer if you feel threatened.

3. Search the web for the warning or issue that you carefully wrote down.

4. Carefully follow the procedure(s) you found to resolve the issue.

5. If you are still perplexed, get some *good help*.

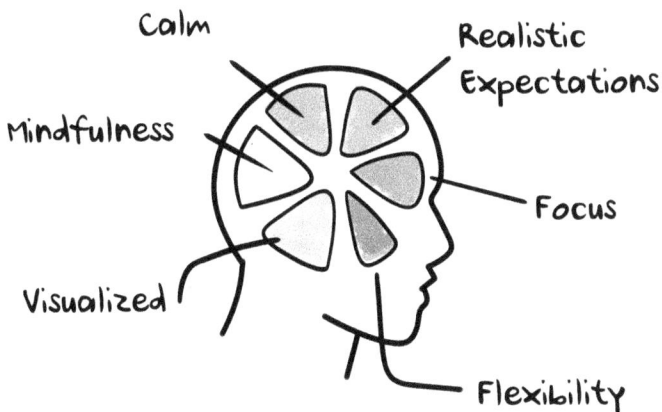

Successful computer use begins, continues and ends with what is inside your head.

Hey, There's More!

There are more Help for the Technology Shy books, tips, tutorials, resources, stories and how-to videos for everything we cover in this book.

Point your computer's web browser to www.technologyshy.com.

You could also visit your local library and/or adult community center and ask for assistance in getting to our humble abode on the Web.

Visit
www.technologyshy.com

Once you're there, please feel free to reach out to us with any questions or comments you may have.

And by the way, thank you for reading this book! We sincerely hope you found it to be helpful.

Index

special function keys), 60 (esc key), 64-65 (on forms), 66 (file selected), 67 (text in document selected), 70-71 (web search), 72 (computer search box), 73 (look brand names and logos), 74 (ports), 75 ("About" window), 76 (computer search box), 79 (application graphical user interface), 86 (attachments 📎), 91-96 (think about what you want to do and then look around for a way to do it), 107-108 (factory reset), 121 (support forum procedure), 123 (in "Save as" window), 127

Web form menu, 32, 37

Web search examples, 51, 58 (for keyboard shortcuts), 61, 63, 70-72, 77, 84, 104 (redirected to ads?), 108 (for factory reset), 115 (for customer support phone number), 118 (to remove bloatware), 121 (for support

"End of the Day"

www.ingramcontent.com/pod-product-compliance
Lightning Source LLC
Chambersburg PA
CBHW071424210326
41597CB00020B/3646